BEAUTIFUL BIOMES

DESERT BIOME

by Elizabeth Andrews

Cody Koala
An Imprint of Pop!
popbooksonline.com

abdobooks.com
Published by Pop!, a division of ABDO, PO Box 398166, Minneapolis, Minnesota 55166. Copyright ©2022 by Abdo Consulting Group, Inc. International copyrights reserved in all countries. No part of this book may be reproduced in any form without written permission from the publisher. Cody Koala™ is a trademark and logo of Pop!.

Printed in the United States of America, North Mankato, Minnesota

102021
012022

THIS BOOK CONTAINS RECYCLED MATERIALS

Cover Photo: Shutterstock Images
Interior Photos: Shutterstock Images, 1, 5 (top) (bottom left) (bottom center), 6–7, 10, 11, 15 (right plant), 17,19, 20; diegograndi/Getty Images, 9; Erlantz Pérez Rodríguez/Getty Images, 12; skalapendra/Getty Images, 15 (left plant)

Editor: Tyler Gieseke
Series Designer: Laura Graphenteen

Library of Congress Control Number: 2021942251
Publisher's Cataloging-in-Publication Data
Names: Andrews, Elizabeth, author.
Title: Desert biome / by Elizabeth Andrews
Description: Minneapolis, Minnesota : Pop!, 2022 | Series: Beautiful biomes | Includes online resources and index.
Identifiers: ISBN 9781098241001 (lib. bdg.) | ISBN 9781098241704 (ebook)
Subjects: LCSH: Deserts--Juvenile literature. | Biotic communities--Juvenile literature. | Habitats--Juvenile literature. | Life zones--Juvenile literature. | Desert animals--Juvenile literature. | Desert plants--Juvenile literature. | Desert ecology--Juvenile literature.
Classification: DDC 577.54--dc23

Hello! My name is

Cody Koala

Pop open this book and you'll find QR codes like this one, loaded with information, so you can learn even more!

Scan this code* and others like it while you read, or visit the website below to make this book pop.

popbooksonline.com/desert-biome

*Scanning QR codes requires a web-enabled smart device with a QR code reader app and a camera.

Table of Contents

Chapter 1
The Driest Biome 4

Chapter 2
The Four Deserts 8

Chapter 3
Desert Plants 14

Chapter 4
Desert Animals 18

Making Connections 22
Glossary 23
Index 24
Online Resources 24

Chapter 1

The Driest Biome

A biome is a large, natural area. It is known for the plants and animals that live there, and its **climate**.

desert

freshwater

forest

Watch a video here!

Deserts are the driest biome. They cover 20 percent of Earth. To be a desert, the amount of water

that dries up must be more than the amount of rain that falls each year.

Chapter 2

The Four Deserts

Desert biomes can be broken down into four types. Hot and dry deserts have very warm weather all year. Rain is rare.

Semiarid deserts are also very hot. But they have seasons. Summers there are long and dry.

The continent of Antarctica is a desert!

Cold deserts have long, freezing winters. They even get some snow.

Coastal deserts sit along oceans. These deserts are more **humid** because fog rolls in from the water. They still get very little rainfall, but seasons are more clear.

Chapter 3

Desert Plants

Desert plants are special. They have **adapted** to the harsh **climate**. Some have short roots that quickly absorb any rain. Other plants have long roots that use water stored deep underground.

Plants like cacti have thick, waxy skin. This saves water and **reflects** heat. Cacti also have spikes to keep animals from stealing their stored water.

A cactus called prickly pear grows pretty, pink flowers and fruits.

Chapter 4

Desert Animals

Animals that live in the desert have also **adapted**. Many live underground to stay cool. They come out at night to look for food. Most will get all the water they need from their food.

Camel humps store fat. They use it when food is hard to find.

Complete an activity here!

Reptiles and insects can survive well in the desert. Not many **mammals** live there. The ones that do have big ears to hear **prey** underground. They might also have furry paws to protect against the hot sand.

Making Connections

Text-to-Self

What sort of adaptions would you have to make if you lived in a desert?

Text-to-Text

Have you read a book about any other biomes? If so, how are they similar to and different from deserts?

Text-to-World

Do you know anything about the people who live in deserts? What do you think their lives would look like?

Glossary

adapt – to change something so that it is easier to live in a place.

climate – weather conditions that are usual in an area over a long period of time.

humid – having a high amount of moisture in the air.

mammal – an animal that makes milk to feed its young and usually has hair or fur on its skin.

prey – an animal that is hunted and eaten by another animal.

reflect – to throw back light instead of take it in.

Index

adapt, 14, 18

animals, 4, 17, 18, 21

climate, 4, 14

plants, 4, 14, 17

seasons, 10–11, 13

water, 6–8, 11, 13, 14–15, 17–18

weather, 8, 10–11, 13

Online Resources

popbooksonline.com

Thanks for reading this Cody Koala book!

Scan this code* and others like it in this book, or visit the website below to make this book pop!

popbooksonline.com/desert-biome

*Scanning QR codes requires a web-enabled smart device with a QR code reader app and a camera.